WORLD'S LONGEST-LIVING ANIMALS
500-YEAR-OLD CLAMS!

By Joni Kelly

Gareth Stevens
PUBLISHING

Please visit our website, www.garethstevens.com. For a free color catalog of all our high-quality books, call toll free 1-800-542-2595 or fax 1-877-542-2596.

Cataloging-in-Publication Data

Names: Kelly, Joni.
Title: 500-year-old clams! / Joni Kelly.
Description: New York : Gareth Stevens Publishing, 2019. | Series: World's longest-living animals | Includes index.
Identifiers: LCCN ISBN 9781538216835 (pbk.) | ISBN 9781538216828 (library bound) | ISBN 9781538216842 (6 pack)
Subjects: LCSH: Clams–Juvenile literature.
Classification: LCC QL430.6 K45 2019 | DDC 594'.4–dc23

Published in 2019 by
Gareth Stevens Publishing
111 East 14th Street, Suite 349
New York, NY 10003

Copyright © 2019 Gareth Stevens Publishing

Designer: Andrea Davison-Bartolotta and Laura Bowen
Editor: Joan Stoltman

Photo credits: Cover, p. 1 Andrew J Martinez/Science Source/Getty Images; pp. 2–24 (background) Dmitrieva Olga/Shutterstock.com; p. 5 (top left) SnBence/Shutterstock.com; p. 5 (top right) Seaphotoart/Shutterstock.com; p. 5 (bottom left) AlessandroZocc/Shutterstock.com; p. 5 (bottom right) Peter Leahy/Shutterstock.com; p. 7 (closed clam) BW Folsom/Shutterstock.com; p. 7 (opened clam) Lev Kropotov/Shutterstock.com; p. 7 (background) scubaluna/Shutterstock.com; p. 9 (main) Andreea Acaprăriței/Shutterstock.com; p. 9 (inset) Universal History Archive/Universal Images Group/Getty Images; p. 11 (top) Fred de Noyelle/CorbisDocumentary/Getty Images; p. 11 (bottom) Kelly Bayliss/Shutterstock.com; p. 13 (main) canadastock/Shutterstock.com; p. 13 (inset) Leosls/Wikimedia Commons; p. 17 (main) Helene Munson/Shutterstock.com; p. 17 (inset) karins/Shutterstock.com; p. 19 Josve05a/Wikimedia Commons; p. 21 Nicole Duplaix/National Geographic Magazines/Getty Images.

All rights reserved. No part of this book may be reproduced in any form without permission in writing from the publisher, except by a reviewer.

Printed in the United States of America

CPSIA compliance information: Batch #CS18GS: For further information contact Gareth Stevens, New York, New York at 1-800-542-2595.

CONTENTS

Weird but Wonderful 4

Ming's the King! 8

How Old Are You? 10

Ming's Not 405? 12

Other Old Clams 14

How Are You So Old? 18

More to Learn! 20

Glossary 22

For More Information 23

Index 24

Boldface words appear in the glossary.

Weird but Wonderful

They don't have heads. They don't have backbones. They can be found in any kind of water—hot or cold, salt water or **freshwater**, deep ocean or sandy shore. One kind of this animal can even live over 500 years. We're talking about clams!

5

All clams have two shells. Their shells are held together with a **hinge**. The soft body of a clam—which has its heart, mouth, stomach, and more—is inside its shell. There are over 12,000 species, or kinds, of clams!

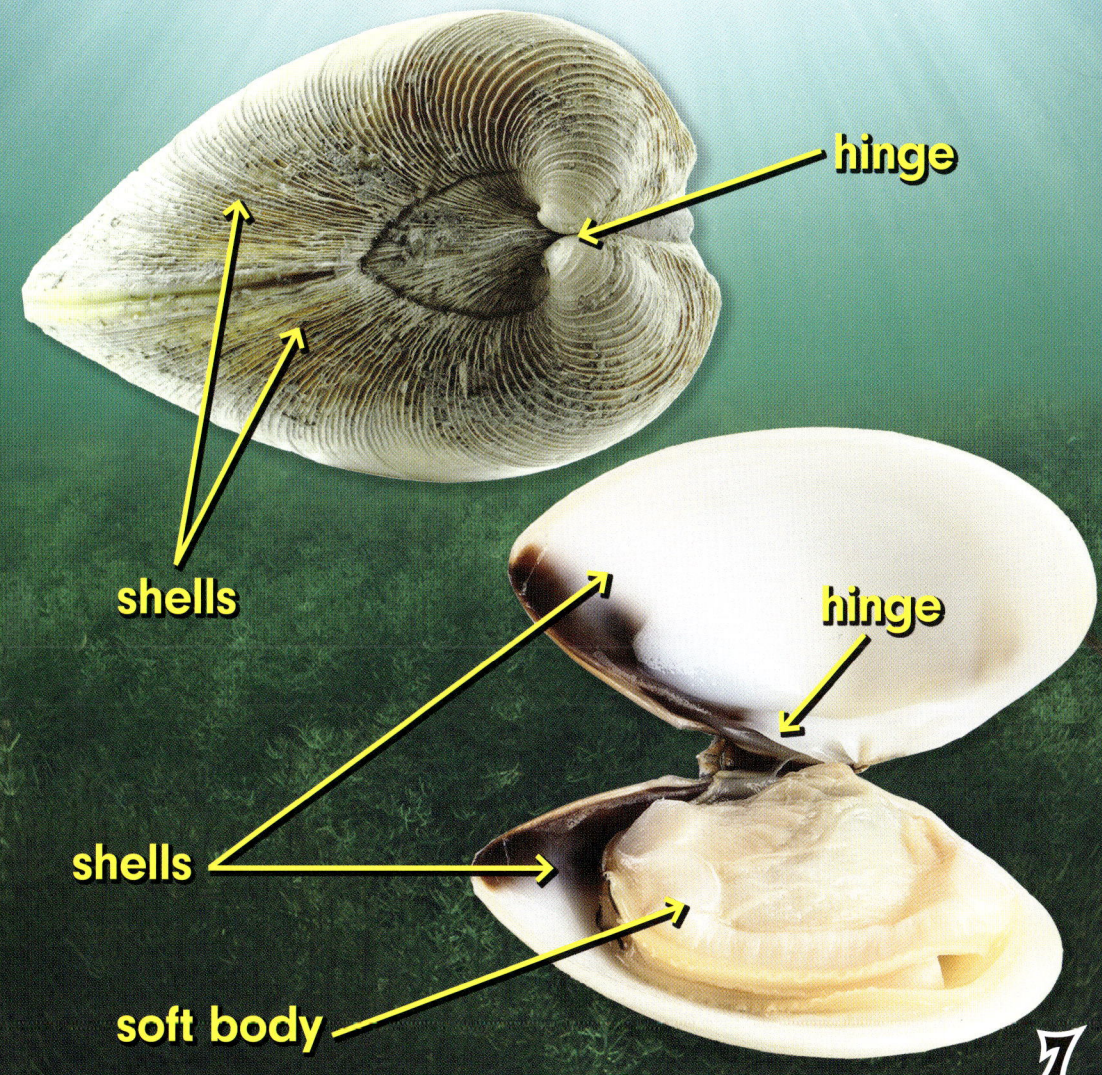

Ming's the King!

In 2006, a group of **scientists** were studying the ocean floor near Iceland. They gathered 200 clams to study. They found that one clam was 405 years old! They named it Ming after the family that ruled China when the clam was born.

CHONGZHEN, THE LAST MING RULER 1627 - 1644

How Old Are You?

A clam's shell grows a new ring every summer. A warm summer with lots of food means a wide ring! Scientists count its rings to see how many summers a clam has lived. They can tell how old a tree is by counting its rings, too!

SHELL RINGS

TREE RINGS

Ming's Not 405?

Clams stop growing as adults, but they keep adding rings to their shell. The rings become **compressed**, with 500 rings in less than 0.1 inch (2.5 mm). Scientists used a special **microscope** to study Ming again. They learned that it was actually 507 years old!

Other Old Clams

Some clams only live a year. Other kinds, like ocean quahogs (KWOH-hahgs), can live a long time! Ming was an ocean quahog. A 220-year-old ocean quahog was found in 1982 off the coast of North America. Another ocean quahog found in 1968 was 374 years old.

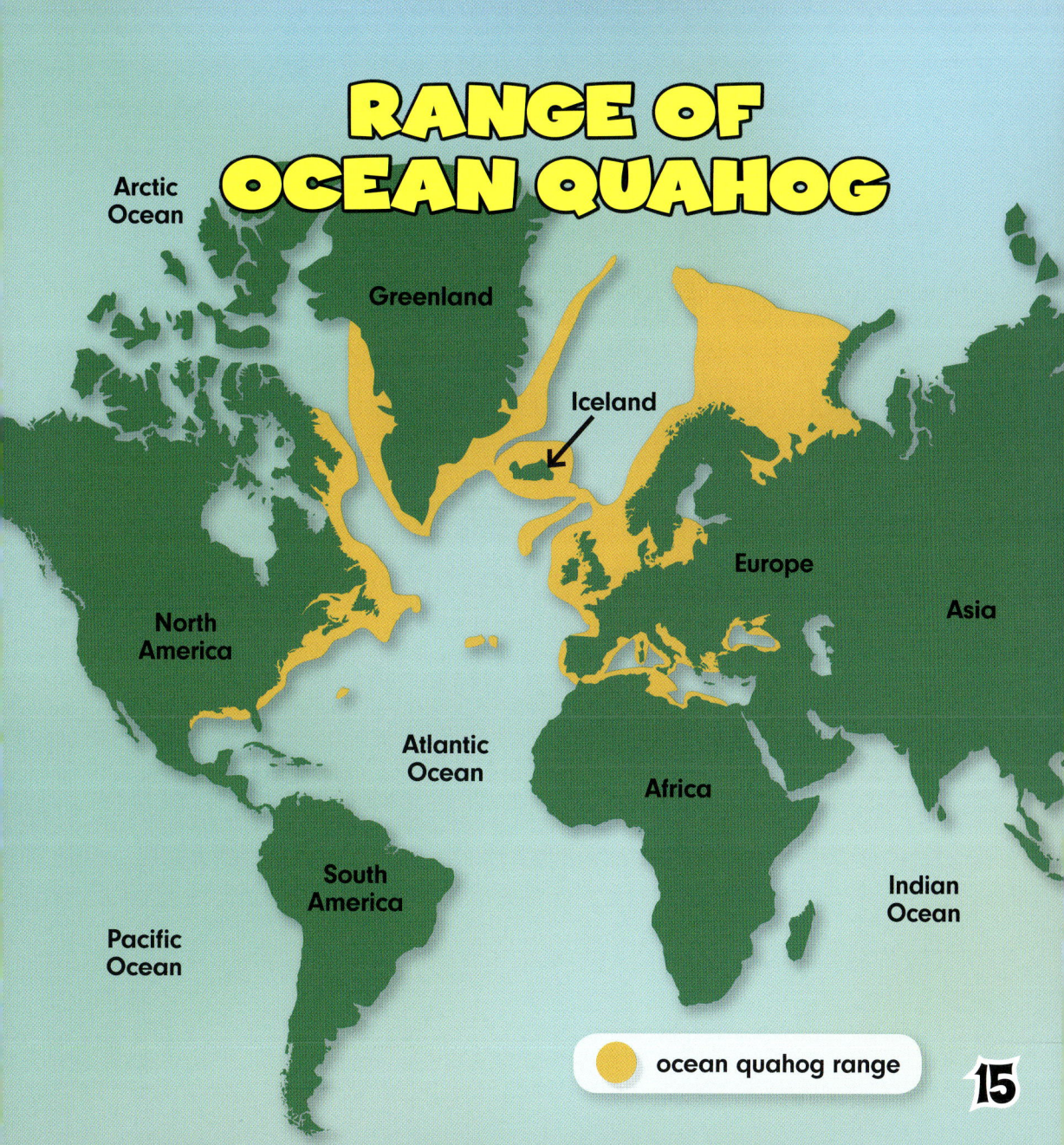

When Ming was pulled out of the water, he died. The scientists wouldn't have touched him if they'd known how special he was. He might have lived much longer! Scientists think there are many old ocean quahogs, though, including some that have ended up in soup.

How Are You So Old?

Ocean quahogs may live so long partly because of their cold **environment**. Many creatures in the deep, cold ocean live a long time, including 400-year-old Greenland sharks! Ocean quahogs also use very little **energy** because they only move a little and have slow life **processes**.

More to Learn!

Clams are very important to their environment. They clean the water and are food for otters, seabirds, and many others. Ocean quahogs are especially important. Studying this special species may teach scientists how aging works!

GLOSSARY

compress: to press or squeeze something so that it is smaller or fills less space

energy: power used to do work

environment: the conditions that surround a living thing and affect the way it lives

freshwater: water that is not salty

hinge: a place where two things connect that allows them to open and close like a door

microscope: a tool used to view very small objects so they can be seen much larger and more clearly

process: a series of changes that happen naturally

scientist: someone who studies the way things work and the way things are

FOR MORE INFORMATION

BOOKS
Machajewski, Sarah. *What Are Mollusks?* New York, NY: Britannica Educational Publishing, 2017.

Metz, Lorijo. *Discovering Clams.* New York, NY: PowerKids Press, 2012.

Young, Karen Romano. *No Bones!* New York, NY: Penguin Young Readers, 2016.

WEBSITES
Clams: Not Just for Chowder!
video.nationalgeographic.com/video/ray_cownosed_eats_clam
Watch this awesome video of a clam on the move!

Meet the Mollusks
biology4kids.com/files/invert_clamsnail.html
Read this fun website to find out more about clams.

What's a Clam?
mass.gov/eea/agencies/dfg/dmf/education/whats-a-clam.pdf
Play games about clams, and read all about them, too.

Publisher's note to educators and parents: Our editors have carefully reviewed these websites to ensure that they are suitable for students. Many websites change frequently, however, and we cannot guarantee that a site's future contents will continue to meet our high standards of quality and educational value. Be advised that students should be closely supervised whenever they access the Internet.

INDEX

clam chowder 16, 17

hinge 6, 7

home 4, 18

Iceland 8, 13, 15

Ming 8, 12, 13, 14, 16

ocean quahog 14, 16, 18, 20

range 15

rings 10, 11, 12

scientists 8, 10, 12, 16, 20

shells 6, 7, 10, 11, 12

soft body 6, 7

tree rings 10, 11